Notes to Big Kids and Grown-ups

1. This is a book about death and contains depictions of dead organisms. Please preview before reading to others so you can take whatever time you need to feel comfortable and prepared.
2. <u>With permission of this book's owner</u>, you are invited to color, scribble, write in, and engage with this book in whatever way best supports you. It is black and white so you can add your own colors.
3. This is an introduction to several complex topics that have been significantly simplified because this is a starting point for further exploration.
4. More abstract topics are visually represented with black and white linework without shading to indicate a much more simplified approach to explaining the concept.
5. There is a glossary of terms that are in bolded text and a library list to continue your learning adventure. None of us have all the answers, including this author, so go learn from the experts with the help of your local library!

May this book offer a helpful tool for you and your loved ones in preparing for or processing conversations on death and dying. For more resources, this author suggests visiting your local library and OrderOfTheGoodDeath.com

Our Stardust

written and illustrated by Karen Daily

For Liam,

my biggest inspiration and supporter.

With eternal gratitude to the many family and friends who shared their thoughtful feedback and kept me going when my motivation waned.

"Look – the Cygnus constellation!"
exclaimed Mini.

"Oooh, it's beautiful" marveled Max.

"Hey, Max?" Mini asked hesitantly.

"Yes, Mini?" Max replied.

"I have a big question…"

"What's your question?"

Mini stared up at the stars and asked,
"What happens when we die?"

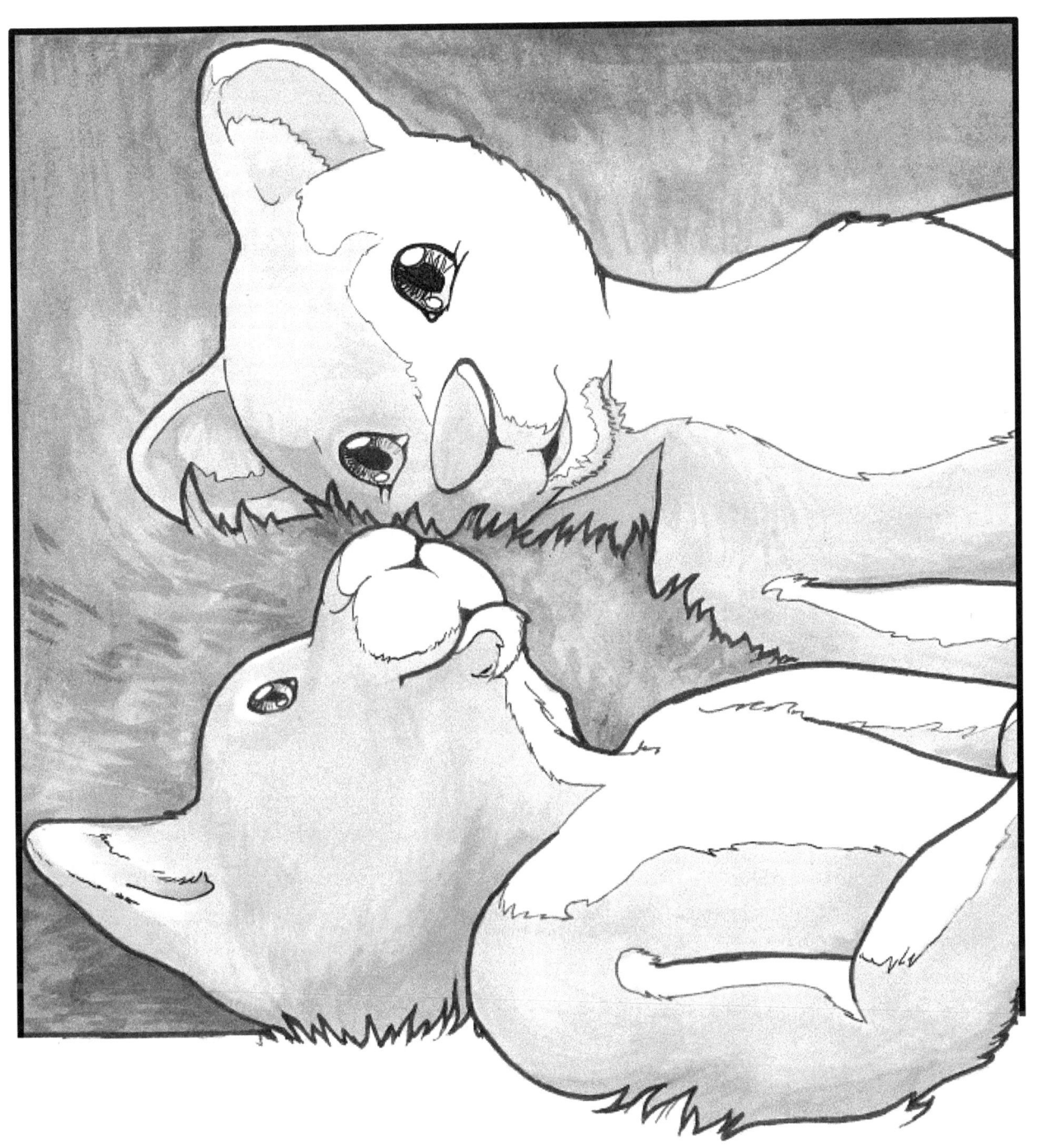

Max turned to face Mini and said,

"That is a big question! It is also a very good question!"

Max continued, "Before we start, this is a really big topic and it is normal to feel big feelings, especially if someone important to you has died."

"We can take as many breaks as you need."

"When you are ready,
take my paw and turn the page."

"There are many questions about death, and we are still learning," Max begins, "but here is some of what we know about the physical part."

"To talk about what happens NEXT, we need to understand a few things from BEFORE and NOW…"

"Everything – including you and me – is made of very teeny, tiny things called **atoms**."

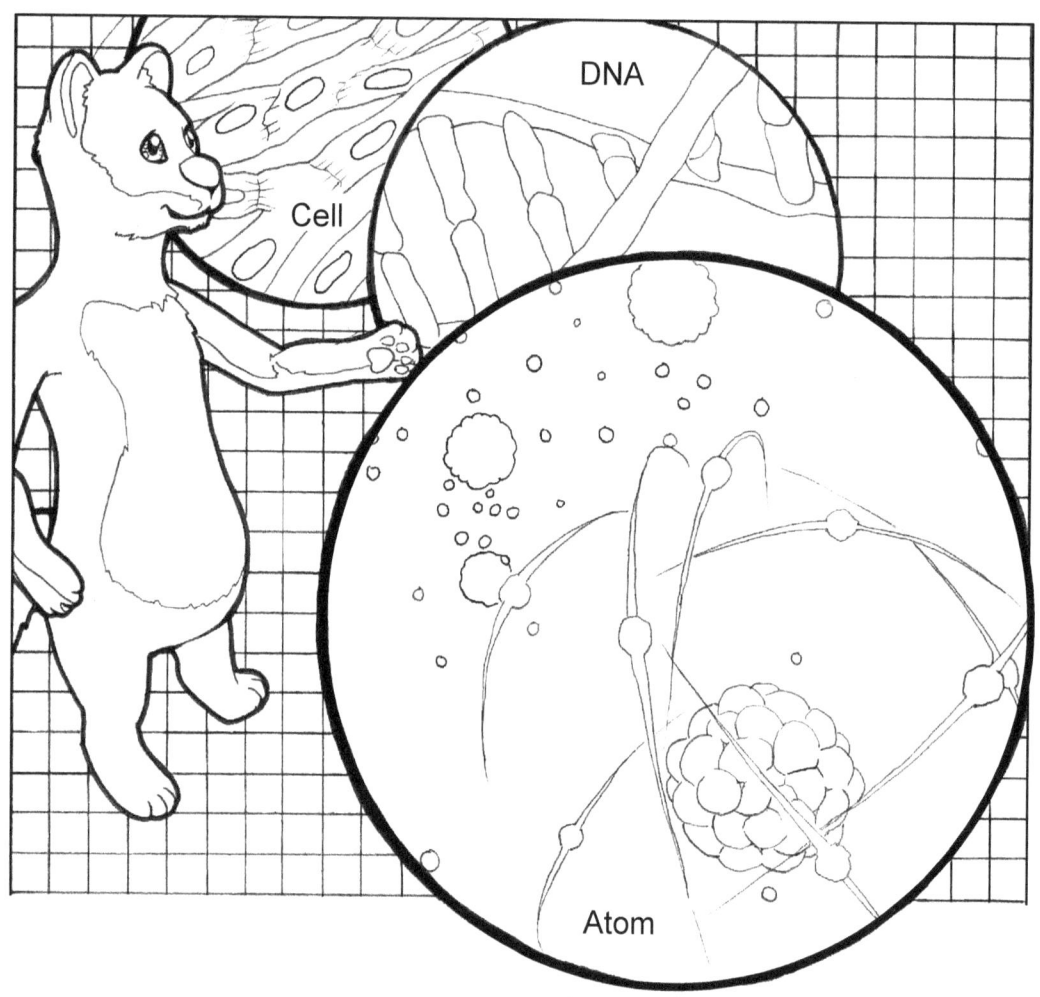

"These atoms are all believed to have originated from the beginning of the universe, the earliest stars, and the dust they left behind when they died."

"Stars, with the power of gravity, squeeze atoms together, creating bigger atoms and a LOT of energy. When a star is no longer able to do that, it dies and leaves a cloud of dust behind. The stardust."

"Gravity pulls all of that STARDUST from dead stars together to form new stars, sometimes with planets!"

"Planets like Earth?" Mini asked.

"Exactly!" Max replied

Stardust doesn't disappear – it becomes part of something new!

The same stardust that formed Earth is in all living things on Earth now. Living things like you and me!

"We are part of that same stardust cycle right now!" exclaimed Max. "Every time we sweat and that sweat evaporates, we are giving back to the **water cycle**."

Mini wonders, "What other examples are there for how we get or give stardust in life?"

There are many ways stardust can return
to this cycle when a living thing dies.

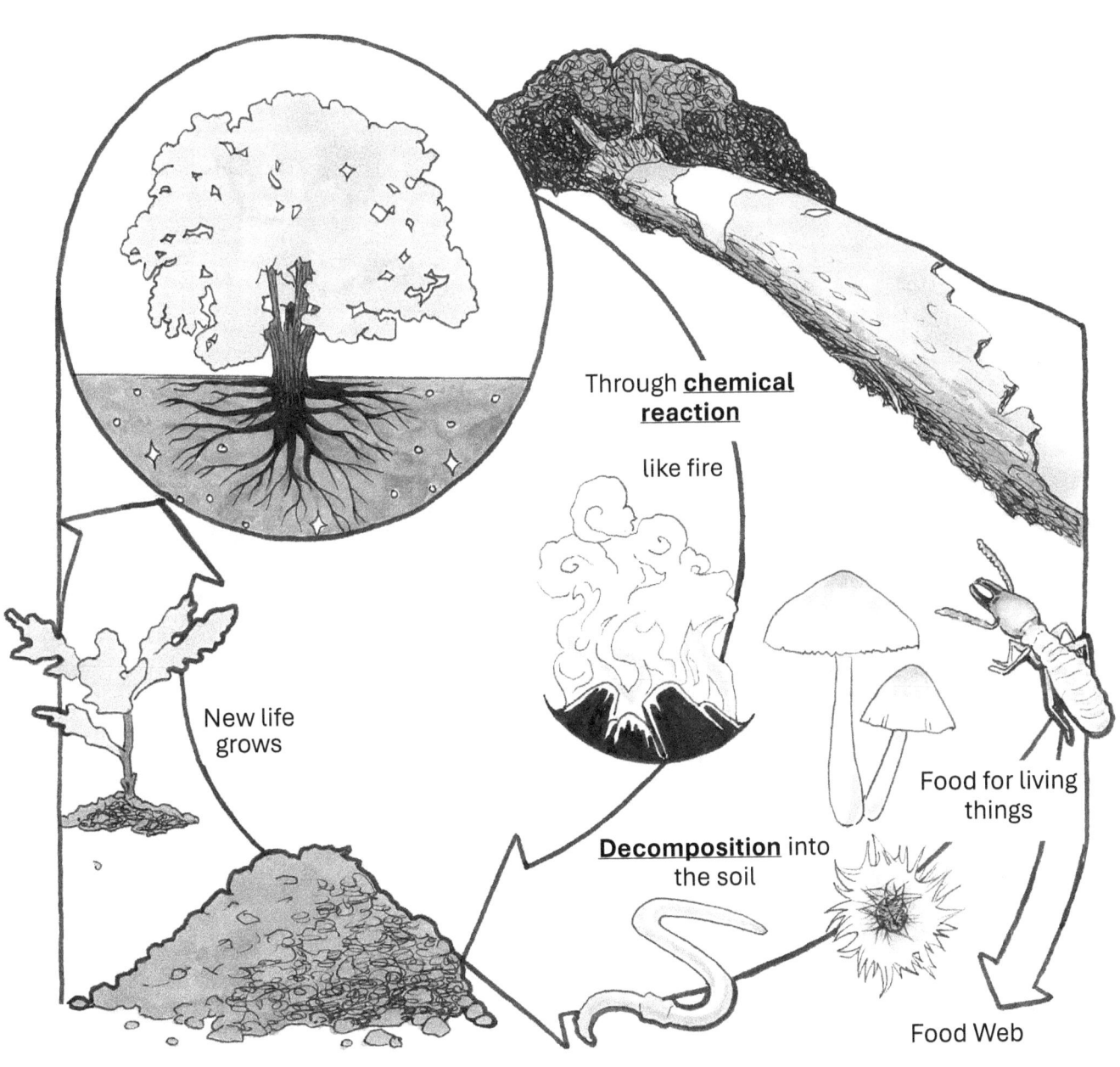

Through **chemical reaction**

like fire

New life
grows

Decomposition into
the soil

Food for living
things

Food Web

Max and Mini looked down to the creek,
past the downed tree and lotus blossoms,
to where the salmon swam.

"When a living thing dies," Max explained,
"it goes through changes all the way down
to the stardust level and that stardust
returns to the cycle as something new."

Mini asks, "How many ways is stardust moving through the cycle here?"

"So, when someone dies, their stardust stays in the world?" Mini asks.

"Yes," Max replies.

"We can't touch, talk to, or visit with them in the same way as when they were alive, but their stardust continues in new ways."

Mini gazed at a dew-covered spider web
and pondered, "What about souls and
spirits and stuff?"

"Great question to research more!" Max smiled. "Curiosity and asking 'why?' help us explore and learn more than what we know now."

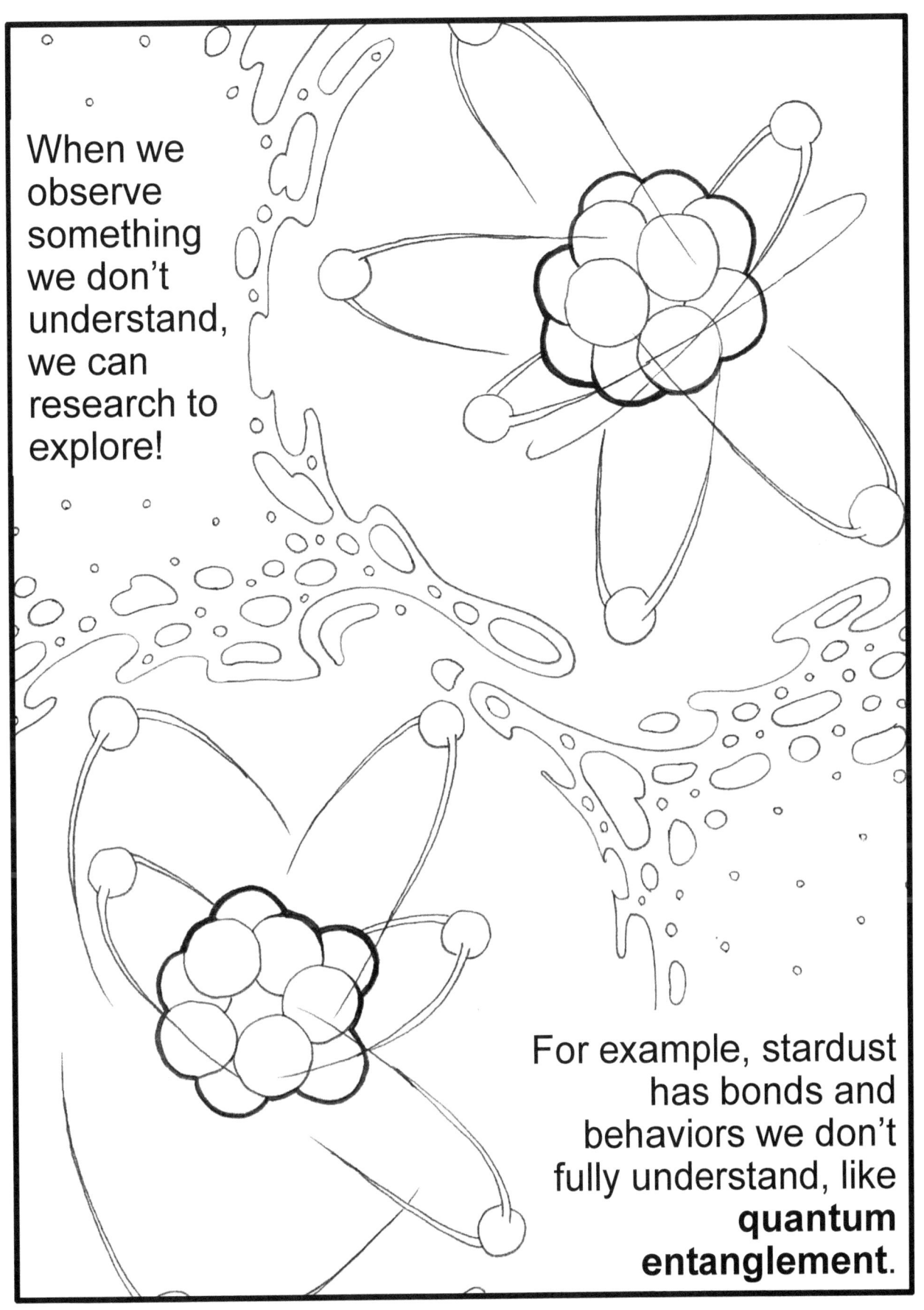

When we observe something we don't understand, we can research to explore!

For example, stardust has bonds and behaviors we don't fully understand, like **quantum entanglement**.

Some of the things we now know to be true were thought to be too weird to be true until we learned more. So, if someone sees, hears, or feels something that seems connected to someone who died, like a special visit from a butterfly, there could be more to learn! Stay curious!

Mini thought for a moment then said, "I still have a lot of questions and big feelings about this."

"That is very normal, and I do, too!" Max exclaimed. "Should we make a plan to go to the library to explore more?"

"Yes, please!" Mini replied.

Library List

Questions I want to research

1. _____

2. _____

3. _____

4. _____

5. _____

6. _____

7. _____

8. _____

Glossary

(Each of these words has so much more to learn!)

Atoms	The basic building block of all matter, made of protons (positively charged particles) and neutrons (neutral particles) in the nucleus and circled by electrons (negatively charged particles).
Chemical Reaction	Reactants, the starting set of molecules (group of two or more atoms), disassemble and reassemble into products, the ending set of molecules with different properties from the starting set.
Constellation	A grouping of stars that, if we connect the "dots" (stars), form something of a picture, often an animal or character, with stories and meanings associated with them.
Decomposition	The process by which something bigger gets broken down into component parts. This can apply to many things, including challenges that need to be tackled and the matter that makes up living organisms (ex. Bacteria, fungi, plants, and animals)
Quantum Entanglement	Special particles that have been separated and, despite the distance between them, still seem to share a bond that shows when the state of one particle is measured, it is known what the state of the other particle would be in that same moment.
Water Cycle	The ongoing cycle of water through Earth and the atmosphere. This includes evaporation (liquid water to vapor), transpiration (vapor released by plants), condensation (vapor forms water droplets in clouds), precipitation (ex. rain drops, snow, and hail), and runoff (water moving downhill to larger bodies of water).